AQUATIC ECOSYSTEMS

Coral Reefs

John Willis

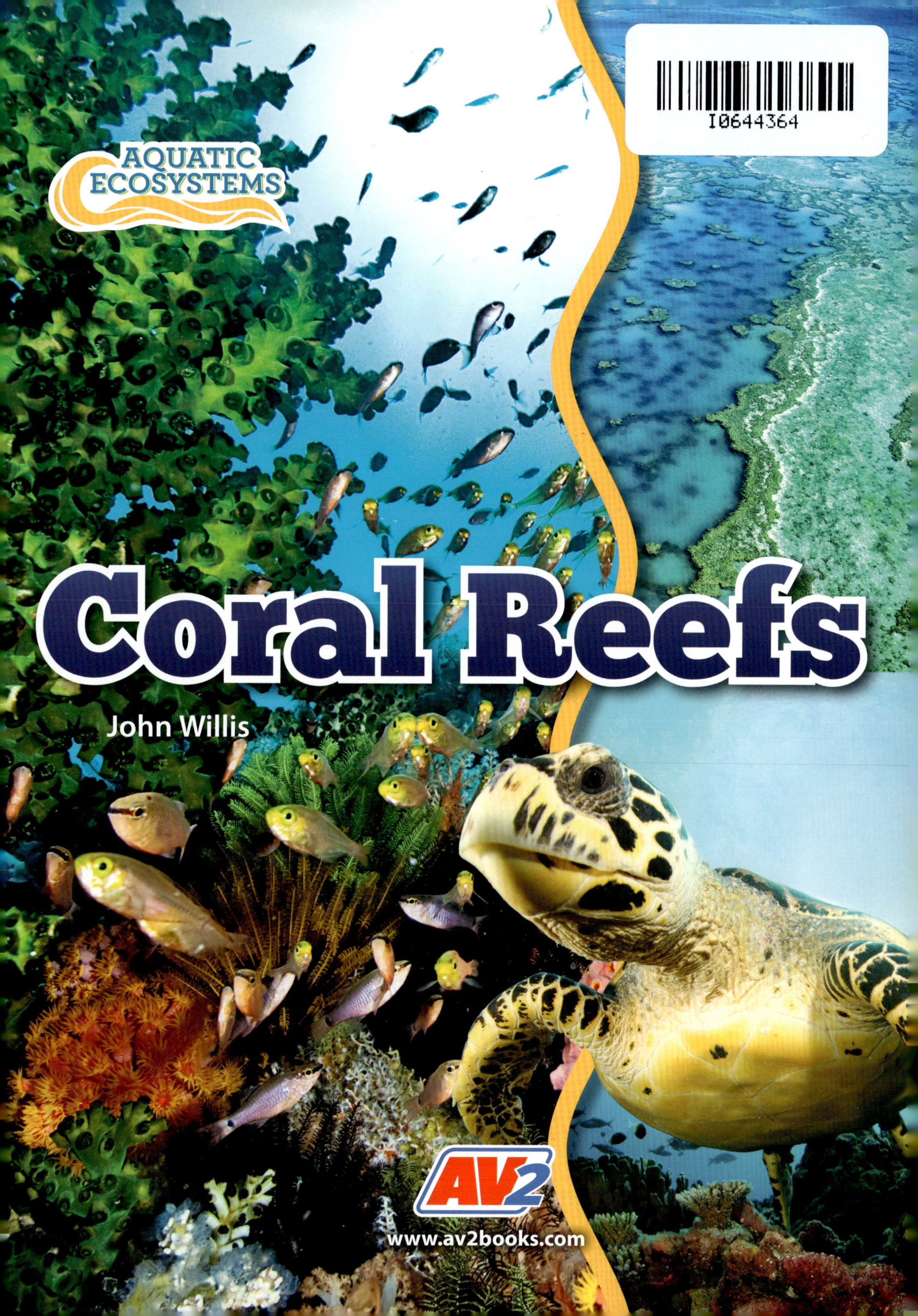

AV2

www.av2books.com

Step 1
Go to www.av2books.com

Step 2
Enter this unique code

XOGPCFHLG

Step 3
Explore your interactive eBook!

AV2 is optimized for use on any device

Your interactive eBook comes with...

Contents
Browse a live contents page to easily navigate through resources

Audio
Listen to sections of the book read aloud

Videos
Watch informative video clips

Weblinks
Gain additional information for research

Try This!
Complete activities and hands-on experiments

Key Words
Study vocabulary, and complete a matching word activity

Quizzes
Test your knowledge

Slideshows
View images and captions

... and much, much more!

AQUATIC ECOSYSTEMS

Coral Reefs

Contents

Remarkable Reefs

Many human **settlements** have been built on coasts around the world. These locations may be vulnerable to different natural disasters such as floods and **tsunamis**. Coral reefs only cover a tiny amount of the ocean's surface, but they play a vital role in protecting coasts from these natural disasters.

Coral reefs are also very important **ecosystems**. They are home to more than 25 percent of all living things in the ocean. Millions of different species of plants and animals live in them. Additionally, young fish of species that live in the open ocean often use coral reefs as shelter until they are large enough to survive on their own. Because coral reefs are so full of life, they provide a great deal of food for many animals. Humans also obtain large amounts of food from coral reef fish and other animals.

Record-Setting Reefs

Oldest Fossilized Reef

Chazy Reef, a **fossilized** reef in **Vermont**, is more than **460 million** years old.

Southernmost Coral Reef

Coral reefs off **Lord Howe Island,** in the Australian territory of New South Wales, are **farther south** than any other coral reef on Earth.

Largest Reef

Australia's Great Barrier Reef is so **large** that it can **be seen** from **space.**

Coral survives best in warm waters. Most of the world's coral reefs are found in tropical locations.

What Are Coral Reefs?

A coral reef is an underwater structure made from the skeletons of tiny **organisms** known as coral polyps. These soft-bodied creatures are related to other ocean animals, such as jellyfish or anemones.

Coral polyps create their hard, rocklike skeleton out of a substance called calcium carbonate that is in the water around them. New polyps create their skeletons on the skeletons left by the polyps that came before them. Over time, this can cause massive coral reefs to be created.

Coral polyps are transparent, but they have many colored **algae** living within their bodies. These algae produce energy using sunlight, in a process called **photosynthesis**. The polyps then use this energy for nutrients. In order to allow their algae to produce energy, coral polyps must live near the water's surface, where sunlight can easily reach them.

Atolls are often close to or below the surface of the water. This can make them dangerous to passing ships.

There are four different types of coral reefs. Fringing reefs grow near shores. They are separated from the shore by thin **lagoons**. Most coral reefs are fringing reefs. Barrier reefs are found much farther from shores. Patch reefs form on the open seafloor. Atolls are also found in the open sea. They are made of coral growing on top of underwater mountains or sunken islands.

Barrier reefs often contain channels. These are passages connecting their lagoon to the surrounding ocean.

Longest Coral Reefs

Great Barrier Reef
Australia
1,553 miles
(2,500 kilometers)

Red Sea Coral Reef
The Red Sea
1,180 miles
(1,900 km)

New Caledonia Barrier Reef
New Caledonia
932 miles
(1,500 km)

Mesoamerican Barrier Reef
Mexico
585 miles
(941 km)

Florida Reef
United States
360 miles
(579 km)

Shaping the Earth

Earth's oceans can affect nearby land in many ways. Waves wear away large areas of coastline in a process called erosion. This process may take place over the course of many years and can greatly affect the ability of humans to live in an area. This is because the foundations of buildings may be lost when coastlines wear away. Other ocean forces can have an even more dramatic effect. Large waves known as tsunamis can cause massive floods. These natural disasters may lead to the destruction of property and the loss of many lives.

Coral reefs serve an important role by preventing or reducing the effects of erosion and natural disasters on Earth's surface. When waves strike a coral reef, they are greatly weakened. They may break before reaching the water's edge. When this happens, very little of the wave's energy is left to have an effect on the shore.

In 2004, a massive tsunami struck countries on the Indian Ocean, including Thailand, India, and Sri Lanka. In total, more than 230,000 people were killed. In some areas, where coral reefs had been destroyed by human activity, the waves caused a large amount of damage. In other locations, where reefs were **intact**, coral may have saved many lives.

The 2004 Indian Ocean tsunami damaged many coral reefs. Although scientists thought it would take decades for them to recover, Indonesia's were able to recover in about four years.

In areas where coral reefs have been destroyed or damaged, people may need to create artificial barriers in order to protect coastlines.

Major Reefs

Coral reefs are found near shores all over the world. Because they are home to so many different plants and animals, they are often known as "rainforests of the sea." Coral reefs are popular tourist destinations. Today, many reefs have been threatened by human activities.

Amazon Reef

In 2016, a new coral reef was discovered at the mouth of the Amazon River in Brazil. The coral reef covers an area of about 3,600 square miles (9,324 sq. km). Unlike many other reefs, the Amazon reef is home to many animals that live in low-light conditions. This is because the water at the mouth of a river is often very muddy.

Florida Key Reef System

The Florida Keys Reef System is the only barrier reef alongside a North American shoreline. It is responsible for about $6 billion of Florida's economy every year. During the last 50 years, the reef system lost about 90 percent of its mass. In 2019, the U.S. government put a plan into place to restore more than 3 million square miles (7.8 million sq. km) of reef.

Great Barrier Reef

The Great Barrier Reef is a well-known tourist destination. As of 2017, more than 64,000 jobs depended on it. In recent years it has been threatened by warming ocean temperatures.

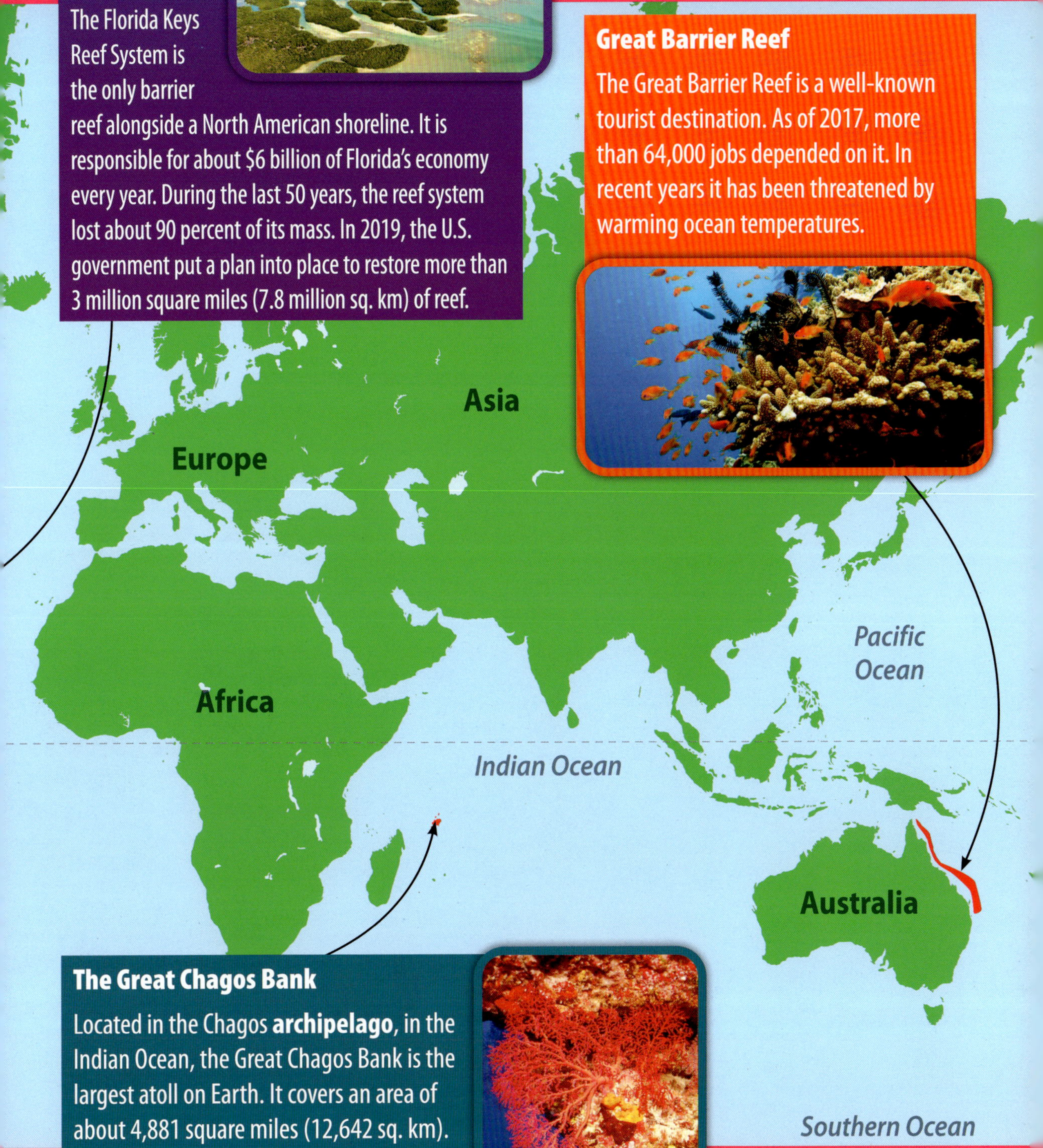

The Great Chagos Bank

Located in the Chagos **archipelago**, in the Indian Ocean, the Great Chagos Bank is the largest atoll on Earth. It covers an area of about 4,881 square miles (12,642 sq. km). While most of this atoll is underwater, small areas rise above the ocean's surface. These **islets** are protected by law.

A Reef Community

All coral reef ecosystems around the world share some common features. In every ecosystem, different organisms occupy certain roles. Energy passes between these plants and animals when they eat or are eaten.

Organisms called producers receive energy from the Sun. They turn this energy into a form that they can use. When other organisms eat producers, they take this energy for themselves. When an organism that eats a producer is eaten, the energy is passed on again. This means that all organisms in an ecosystem need producers for their energy.

Primary consumers are animals that eat producers. Most primary consumers are **herbivores**. **Carnivores** and **omnivores** feed on primary consumers. This is why they are known as secondary consumers. A tertiary consumer is a carnivore or omnivore that has no **predators**. Tertiary consumers get their energy from eating primary or secondary consumers. Finally, when organisms die, **decomposers** break down their remains. This recycles their energy. It turns into nutrients that producers can use.

Some tertiary consumers eat a wide range of different prey. Lemon sharks eat fish, crustaceans, and even seabirds.

Coral Reef Energy Pyramid

As energy passes through an ecosystem, some of it is lost. This is why there are fewer consumers than producers in an ecosystem. This relationship can be shown through an energy pyramid.

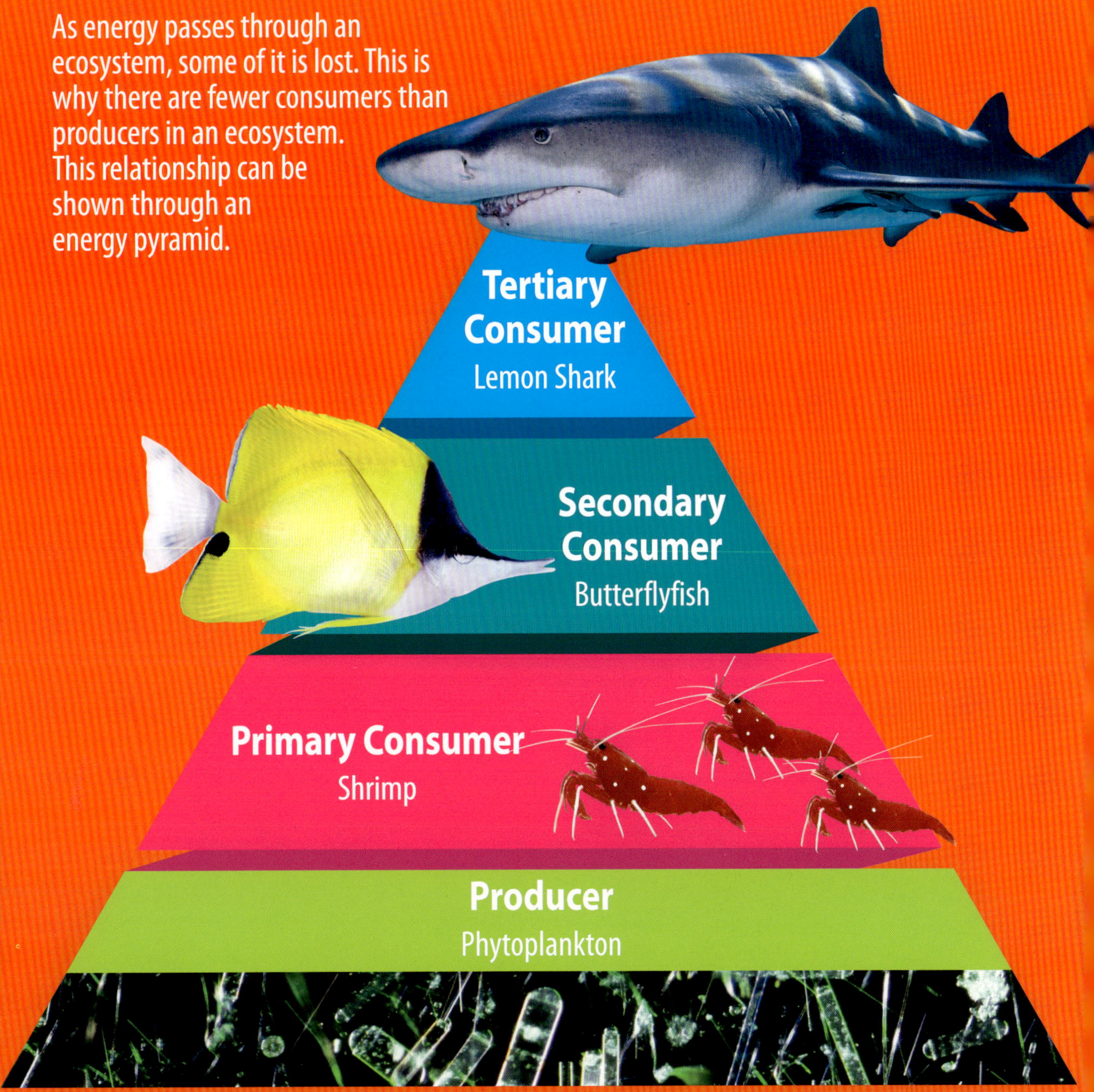

Try This! Choose a coral reef plant or animal. Research it using the library or the internet. Is it a producer or a consumer? If it is a consumer, what kind is it? Use your research to create an energy pyramid that includes your chosen organism.

Coral Reef Producers

Most of the producers in a coral reef ecosystem are plants or algae. Many coral reef algae survive by living with coral polyps. These algae, known as zooxanthellae, are too small to be seen by the human eye. Many organisms eat coral polyps and the algae within them. When this happens, energy from the Sun is passed on to other organisms. When ocean temperatures rise too much, coral polyps lose their algae. Without algae in their bodies, coral polyps have a much more difficult time creating the energy they need to survive. When coral polyps lose their algae, they also become pale or white. This is why this process is known as bleaching.

There are about 72 different species of seagrasses.

Unlike algae, seagrasses are true plants. They use leaves to create energy using photosynthesis. Their roots keep them anchored to the ground. Seagrasses are often found in clear, shallow waters along with coral. This makes it easier for them to obtain the sunlight they need to produce energy. Coral reefs provide many benefits for seagrasses. They help prevent currents and waves from uprooting them or damaging their leaves. Seagrasses help protect reefs as well. They can remove pollution from the water and can take in fresh water that could otherwise damage coral. Seagrasses are also useful for many reef fish. These fish use seagrasses as hiding spots to stay safe from predators.

Zooxanthellae have been living in coral polyps for more than 160 million years. They give coral its vivid colors.

Coral Reef Fish

Coral reefs are home to a vast variety of different fish species. Fish in coral reefs have unique adaptations that help them survive. While fish in rivers and oceans are often built for speed, coral reef fish do not need to move as quickly. Instead, they must be able to maneuver easily and fit through small gaps and crevices. This is why many coral reef fish have flat bodies. Many fish that live in coral reefs have bright, vivid colors. This allows them to hide more easily among the colored coral.

Like other eels, moray eels have long, narrow bodies. This important adaptation allows them to make their homes in small crevices and holes in reefs. Moray eels can feed on smaller fish, crustaceans, and octopuses. An eel waits for fish to swim by its home. Then, it quickly lunges at and grabs its prey with its sharp teeth. Moray eels have a second set of jaws in their mouth. These help them to pull their food down their throats.

Although most moray eels grow to be about 5 feet (1.5 meters) long, the biggest can be 11.5 feet (3.5 m) in length.

Some coral reef fish have highly specialized bodies. Seahorses are some of the slowest-moving fish. The move far too slowly to be able to escape from predators. Instead, they spend most of their time using their tails to cling to coral. Their bodies are brightly colored to match the surrounding coral. Seahorses use their narrow, toothless mouths to feed on tiny animals. They catch prey by sucking it into their mouths.

Lionfish are native to the Indian and Pacific oceans but have been introduced to coral reefs around the world. Lionfish mostly feed on smaller fish. They are poor swimmers, but they do not need to escape from predators. Besides their bright colors, these fish have large, venomous spines in their fins. Because they are so difficult for other animals to eat, this has caused them to become an **invasive species** around the world.

There are about 36 different species of seahorses. The largest are more than 1 foot (30 centimeters) long.

Cold-Blooded Coral Reef Animals

A cold-blooded animal is one that cannot maintain a constant body temperature on its own. Instead, it must rely on its surroundings to cool off or warm itself. These animals are also known as ectotherms. Many fish, along with reptiles, mollusks, and crustaceans, are ectotherms.

Coral Reef Reptiles

Reptiles are cold-blooded animals with scaly bodies. Although many reptiles live in both the water and on land, some spend almost their entire lives in water. Reptiles commonly found in coral reefs include sea turtles. Sea turtles get most of the water they need through the food they eat or by drinking salt water. Special **glands** in their bodies allow these animals to get rid of extra salt.

Sea snakes are another common coral reef reptile. These snakes have paddle-like tails to help them move easily through the water. Some sea snakes can breathe through their skin. This allows them to stay underwater for hours without breathing.

Sea snakes have very strong venom. However, they are not aggressive toward humans.

There are about 100 different species of cuttlefish.

Coral Reef Mollusks

Mollusks are soft-bodied animals that include snails and slugs. Cephalopods, which include squid, cuttlefish, and octopuses, are mollusks that live in the ocean. These animals use their kidneys to filter salt out of the water they take in.

Cuttlefish are able to change the color and texture of their skin in order to match the coral around them. By hiding among the coral, they can ambush small fish that pass by. Like cuttlefish, octopuses can also change their color and texture. This ability, along with their soft bodies, allows them to hide in small spaces throughout coral reefs.

Coral Reef Crustaceans

Crustaceans are hard-shelled animals. Although some live on land, the majority of these animals make their homes in the water. Crustaceans include crabs, lobsters, and shrimp.

Some coral reef shrimp have developed a unique method of feeding. These shrimps set up "cleaning stations" on the reef. When a fish approaches, the shrimp waves its antennae to get the fish's attention. Then, the shrimp will clean the fish. The shrimp gets food, while the fish has **parasites** removed from its body.

Cleaner shrimp clean the skin, gills, and even the mouths of many species of fish.

Warm-Blooded Coral Reef Animals

Hawai'ian monk seals are some of the most endangered marine mammals on Earth.

Warm-blooded animals can maintain a constant body temperature. They are also known as endotherms. Both birds and mammals are endotherms. These animals do not depend on their environment to keep warm or cool. Many of these animals rely on coral reefs for survival.

Coral Reef Mammals

Hawai'ian monk seals spend most of their lives in the water around the Hawai'ian islands. They are some of the only seals to spend their lives in warm waters. These seals eat small animals that they find living under rocks in coral reefs.

Once a year, monk seals molt, or shed their skin. When they do so, they need to find a protected area on land. When these seals give birth, they find a sandy, safe beach.

Whales and dolphins also often depend on coral reefs for food and shelter. More than 30 different whale and dolphin species are found in Australia's Great Barrier Reef. Bottlenose dolphins are found in waters around the world, including in coral reefs. These dolphins eat many kinds of small fish. In some coral reefs, they use tools to hunt. They grab sponges using their mouths. This protects their snouts from coral while they hunt for prey.

Coral Reef Birds

Because most birds can fly, they are often able to live in several different ecosystems. Birds often take advantage of the large amount of fish living in coral reef ecosystems.

Seabirds are important to coral reefs as well. These birds mostly eat small fish that are full of nitrogen. When nitrogen leaves the bird's bodies in their droppings, it enters the water in and around coral polyps,which use it to grow more quickly. The algae that many coral reef fish feed on also depends on nitrogen from birds to grow.

The magnificent frigatebird often nests near coral reefs. When not nesting, these birds spend almost all of their lives in the air. Frigatebirds do not have waterproof feathers. They hunt by taking fish from the water's surface or by stealing from other birds.

Male frigatebirds use a bright red throat patch in order to attract mates.

Use and Overuse

Coral reefs are very important to people. One of the most common uses is as a source of food. Reefs are used for both commercial and recreational fishing. In the United States alone, reefs are worth about $100 million each year solely for the fish caught for food. Coral reefs are important in developing countries as well. In many poorer areas of the world, people depend directly on the fish that they catch from coral reefs for the food that they eat.

In more recent years, reefs have also become very popular tourist destinations. People enjoy snorkeling or diving in these ecosystems and seeing the vast variety of life that they hold. Other people take boat tours to see marine mammals living nearby. Worldwide, coral reef tourism is responsible for about $36 billion each year. Some smaller islands depend almost entirely on tourism for money.

Coral reefs are also important in the field of medicine. Many animals in coral reefs, including the coral themselves, do not move quickly. Instead of fleeing from threats, they have developed chemical defenses to keep safe. Scientists are able to take these chemicals and use them to develop new drugs and medicines. In the future, new treatments for diseases such as cancer may come from discoveries made by examining coral reef life.

It is possible to use the skeletons left behind by coral to replace damaged bones in humans.

Overfishing and Coral Reefs

While fishing in coral reefs is important to many people, overfishing can be a source of danger for coral reefs.

Fishing in Nurseries	Coral reefs are home for many young fish. Fishing in parts of a reef where these fish grow can stop them from being able to reproduce.	
Ghost Traps and Lost Nets	Sometimes, people lose or forget traps and nets. When this happens, they may continue to catch and kill fish for many years.	
Fishing When Fish Spawn	Some fish gather in large groups to mate and lay eggs. If they are caught at this time, far fewer new fish will be born.	
Catching Large Fish	Big fish are more likely to have strong, healthy young. If people only catch the largest fish, it can make it more difficult for the species to survive.	
Non-Selective Traps	Some traps and nets catch all kinds of fish. This means that fish that would not be caught by bait, such as herbivores, are captured. Many other fish depend on these fish for food.	

Think About It! People may need to catch lots of fish for food or to make a living. Can this be balanced with the need to keep reef ecosystems safe? What steps do you think could be taken to find this balance?

Coral Reefs in Danger

Coral reefs today face several threats. The spread of some predatory animals has damaged many reef systems. The crown-of-thorns starfish is an animal that naturally eats coral. In large numbers, these starfish can heavily damage a reef. In the late 1960s, their population increased greatly in Japan and Australia. They began to eat large areas of coral reefs. Scientists believe that nutrients from farming and other human activity had washed into the ocean. Starfish that fed on these nutrients had more chances of survival. Additionally, in some areas, people had taken the snails that fed on them for their shells. This reduced the number of the starfish's predators.

Coral bleaching is one of the biggest threats facing coral reefs today. Typically, bleaching takes place when water temperatures rise above 86° Fahrenheit (30 °Celsius). If the water temperature lowers again, the algae in coral can regrow. The reef will recover from bleaching over time. However, if the temperature does not lower, the coral will begin to starve. Sunlight also can lead to bleaching. Coral in shaded areas last longer in higher temperatures before bleaching.

A reef ecosystem suffers great damage when large areas of coral become bleached. In 2005, half of all U.S. reefs in the Caribbean were lost due to a large bleaching event. This was caused by warm water moving south. Not all bleaching events are caused by warm water. In 2010, a large bleaching event took place in Florida when water temperatures dropped too quickly.

An adult crown-of-thorns starfish has up to 19 arms.

Temperature is not the only cause of coral bleaching. Other causes include chemicals found in some sunscreens.

Protecting Coral Reefs

Because coral reefs are so important to so many people, governments around the world have taken steps to protect them. In 2000, the United States government established the National Oceanic and Atmospheric Administration (NOAA) Coral Reef Conservation Program to study and protect U.S. coral reefs.

Great Barrier Reef Timeline

60,000 Years Ago

The first **indigenous** people reach Australia and encounter the Great Barrier Reef. They begin to use it as a source of food.

1770

Europeans first encounter the Great Barrier Reef when British explorer James Cook runs into it with his ship. He begins charting the reef and the passages through it.

1981–1982

The first known mass bleaching event in the Great Barrier Reef takes place. This event helps people to find the connection between water temperature and coral bleaching.

The U.S. government has also set aside major coral reefs as marine protected areas (MPAs), in which the government monitors and protects the ecosystem.

Internationally, other groups help protect coral reefs as well. The United Nations Environment Programme (UNEP) established the Coral Reef Unit (CRU) to help protect coral reefs around the world, including the Great Barrier Reef. UNEP works with governments to monitor coral reefs and protect them.

The Great Barrier Reef has been greatly affected by coral bleaching. What steps do you think could help to protect it? Who would have to take these steps? Would they cause negative effects for any groups or individuals?

1997–1998

A major mass bleaching event takes place. More than 70 percent of the Great Barrier Reef's inshore reefs are bleached. The cause of the event was the record-setting summer temperatures in the area.

2016

A massive coral bleaching event takes place across the world. More than 20 percent of the Great Barrier Reef's corals are killed. Another mass bleaching event affects the reef again the following year.

2020

In March, the Australian Research Council Centre of Excellence for Coral Reef Studies at James Cook University begins aerial surveys to measure the Great Barrier Reef's health.

Create a Coral Reef Food Web

Plants and animals in a coral reef ecosystem rely on each other for food. This can be modeled using a food web. You can create a food web using the steps below.

1. Find at least three coral reef organisms of each type—producers, primary consumers, secondary consumers, and tertiary consumers.

2. Write down the name of one producer. Then, add the name of a primary consumer that eats it. Use an arrow to show the transfer of energy.

3. Write down the name of a secondary consumer that eats your primary consumer and the name of a tertiary consumer that eats your secondary consumer. Use arrows to show the flow of energy. This will create a food chain.

4. One at a time, add each of the remaining organisms to your food chain. Add new arrows to show which organisms they eat or are eaten by. Once all the organisms have been added, your food chain will be a food web.

Starter Food Web

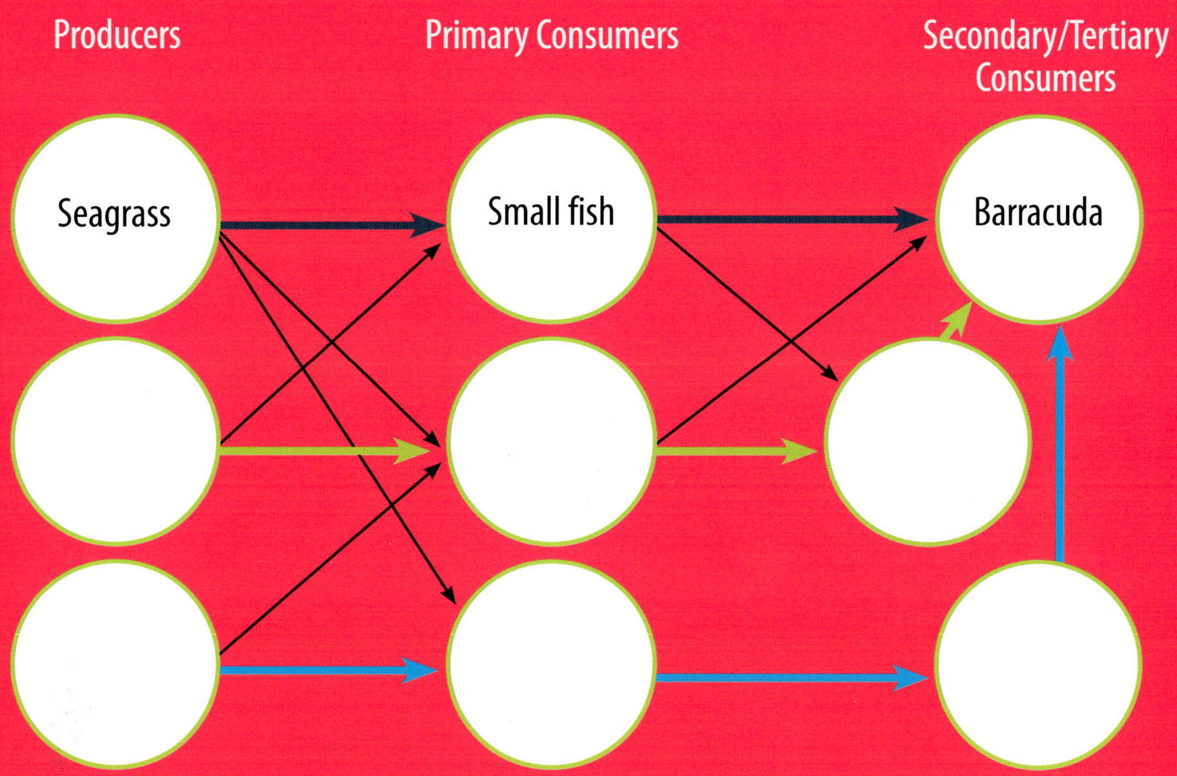

Once your food web is complete, use it to answer the following questions.

1. How would removing one organism from your food web affect the other organisms in the web?
2. What would happen to the rest the food web if the producers were taken away?

Quiz

1 Are lionfish strong swimmers?

2 What happens to waves that strike coral reefs?

3 In which state is the only North American barrier reef?

4 How do cleaner shrimp attract fish?

5 What element from seabird droppings is used by coral to grow?

6 What is a ghost trap?

7 When did Europeans first find the Great Barrier Reef?

8 What can coral skeletons help replace in humans?

ANSWERS

1. No **2.** They are weakened **3.** Florida **4.** By waving their antennae **5.** Nitrogen **6.** A trap that a person has lost or forgotten **7.** 1770 **8.** Damaged bones

Key Words

algae: plantlike organisms that can turn sunlight into energy

archipelago: a group of islands

carnivores: animals that only eat meat

decomposers: organisms that break down dead matter

ecosystems: communities made of living things and the places in which they live

glands: organs that make a substance for the body to use

herbivores: animals that eat plants

indigenous: native to an area

intact: undamaged

invasive species: a plant or animal species that has been placed in a new location and damages it

islets: small islands

lagoons: shallow bodies of water connected to larger ones

omnivores: animals that eat meat and plants

organisms: living things such as plants or animals

parasites: organisms that live in or on another organism and harm it

photosynthesis: a process that produces energy using sunlight

predators: animals that hunt other animals

settlements: permanent communities of people

tsunamis: large waves caused by underwater earthquakes or volcanoes

Index

Published by AV2
14 Penn Plaza, 9th floor
New York, NY 10122
Website: www.av2books.com

Library of Congress Cataloging-in-Publication Data

Names: Willis, John, author.
Title: Coral reefs / John Willis.
Description: New York, NY : AV2, [2020] | Series: Aquatic ecosystems | Includes index. | Audience: Ages 8-14 | Audience: Grades 4-6
Identifiers: LCCN 2020017320 (print) | LCCN 2020017321 (ebook) | ISBN 9781791128074 (library binding) | ISBN 9781791128081 (paperback) | ISBN 9781791128098 | ISBN 9781791128104
Subjects: LCSH: Coral reef ecology--Juvenile literature. | Coral reefs and islands--Juvenile literature.
Classification: LCC QH541.5.C7 W55 2021 (print) | LCC QH541.5.C7 (ebook) | DDC 577.7/89--dc23
LC record available at https://lccn.loc.gov/2020017320
LC ebook record available at https://lccn.loc.gov/2020017321

Printed in Guangzhou, China
1 2 3 4 5 6 7 8 9 0 24 23 22 21 20

052020
101319

Project Coordinator: John Willis
Designer: Ana María Vidal

AV2 acknowledges Getty Images, iStock, Shutterstock, and Wikimedia as its primary image suppliers for this title.